Chemical Engineering & the Law Forum 2020

AA002638

Held at the 2020 AIChE Annual Meeting

Online
16 - 20 November 2020

ISBN: 978-1-7138-2317-9

Printed from e-media with permission by:

Curran Associates, Inc.
57 Morehouse Lane
Red Hook, NY 12571

CURRAN ASSOCIATES INC. proceedings .com

Some format issues inherent in the e-media version may also appear in this print version.

AIChE
120 Wall Street, FL 23
New York, NY 10005-4020

Phone: (800) 242-4363
Fax: (203) 775-5177

www.aiche.org

Additional copies of this publication are available from:

Curran Associates, Inc.
57 Morehouse Lane
Red Hook, NY 12571 USA
Phone: 845-758-0400
Fax: 845-758-2633
Email: curran@proceedings.com
Web: www.proceedings.com

TABLE OF CONTENTS

610545 Intellectual Property and a Sustainable Future

Abstract:

As climate change continues to affect the world, industries of all kind are developing new ways to increase their sustainability and decrease their carbon footprint. This green revolution has led to a variety of new proprietary technologies being brought to the market. With the increase in new sustainable technology, the necessity for intellectual property continues to rise. Although companies need to protect their IP rights, there is also a need to allow other companies to help the world become more sustainable. Thus, companies must balance the need to protect IP rights and the need to share their technologies.

This presentation will provide an overview of the basics of IP, how IP is at the forefront of sustainable technology, and important considerations for members of industry, academics, and anyone with an idea for a green invention. Speakers will also highlight several case studies addressing sustainable IP and whether the IP was strong enough.

616721 How emerging growth companies can effectively protect their intellectual assets

Abstract:

The discussion will include a general overview of intellectual property rights and why they are important for emerging growth companies (e.g., entrepreneurs, start-ups, university spin-outs) in the life sciences sector to consider early in the life of the company. The discussion will particularly focus on how to determine if a patent application is an appropriate means of protection for the emerging growth company and how to effectively pursue a patent application. The discussion will include steps an emerging growth company can perform on their own without having to involve an outside consultant (e.g., attorney). The discussion will include common pitfalls when obtaining intellectual property rights and creative ways emerging growth companies can enforce their rights to deter competition or generate revenue (e.g., licensing).

601618 "Sustainabilty – the Need for Scalable Solutions & a Framework to Make Decisions"

Abstract:

Sustainability had taken a new dimension in recent times with employees and investors challenging industries to be more sustainable. An overview of CO2 generation sources globally and the USA will be presented and a case will be made for the need to scaleable solutions. The scalability and mutual interaction of these solutions have resulted in trade-offs. In this talk, the need for a frame-work for decision making and the role scientists and engineers can play is discussed.

606476 The Impact of Policy Instruments on Deployment of Low-Carbon Heating in the UK: A Whole System Optimisation Study

Abstract:

As part of the UK's commitment to emitting net zero greenhouse gases by the year 2050, the country must decarbonise the space heating sector, which currently contributes 17% of these emissions. This will involve retrofitting 24 million residential buildings, currently using natural gas, to a low carbon form of heating. The main options to achieve this are either hydrogen based, such as hydrogen boilers, or electricity based, such as electric heat pumps and electric resistive heating. Each option involves considerable costs along the supply chain, including upgrading infrastructure and installing new equipment in homes.

One of the key challenges facing policymakers is that, at present, natural gas heating is often the cheapest form available to households. In addition, considerable investment will be required in hydrogen production and transportation before it becomes available to consumers. Consequently, the transition away from natural gas heating is likely to require financial incentives. Each of the strategies is likely to present different costs to different stakeholders and so policies will be required targeting different points of the supply chain to incentivise the uptake of low-carbon heating.

The aim of this study is to assess how policy instruments, designed to promote the uptake of renewable heating, might impact both the heating technology mix and the supporting infrastructure when considering the least cost solution to decarbonisation. The Value Web Model, a mixed integer linear programming model for integrated energy value chains, was developed to optimise the deployment of heating technologies alongside energy generation, transportation, and storage to meet energy demands in the UK. This model considers natural resource availability (including wind, solar, land, and natural gas) alongside energy demand with high spatial and temporal resolution to determine not only which technologies should be used, but where they should be located. The results will show how different incentives applied at different points along the supply chain and in different amounts impact both the penetration of the technology they are aimed at and the overall technology mix. This work will highlight the impacts of incentives on the wider energy system and add to the discussion, relevant to policymakers and researchers, on how to decarbonise energy systems at the lowest cost.

611169 The Role of Chemical Engineering in Sustainable Energy Policy

Abstract:

Technological advancements are often made long before their application is mandated by policy. Conversely, the way a technology is regulated can make or break its adoption. Sustainability of energy involves balancing our energy needs with the well-being of the environment and the community. This talk will examine examples of policy successes and how chemical engineering can contribute to sustainability of energy and energy policy.

600820 Review of US Environmental Protection Agency General Exposure Scenario Modeling for Evaluating Chemical Safety

Abstract:

Generic Scenarios were initially developed by the US Environmental Protection Agency (USEPA) as a part of the chemical risk assessments process required under the Toxic Substances Control Act (TSCA) used by the USEPA to evaluate the hazards associated with the modeled chemical. From these humble beginnings, Generic Scenario documents have evolved into Organisation for Economic Co-operation and Development (OECD) Emissions Scenario Documents (ESDs) produced to describe a chemical's sources, production processes, pathways and use patterns. This presentation provides a review of exposure modeling performed in the development of existing Generic Scenario documents. This review was conducted to understand the evolution of the Generic Scenario process, with a goal of simplifying the modeling approaches used to estimate exposure levels and human health effects related to manufacture and use of chemicals currently required under the Lautenberg Chemical Safety for the 21st Century Act.

597556 Effectively Communicating Sustainability and Engineering within Federal Policy

Abstract:

As engineers and scientists we wonder why it seems that science is void from the decisions made be policy makers and governmental entities. There are scientists and engineers in government, but we have to be willing to put in the work to be effective, and those implementing policy have to be willing to listen. The ideas presented in this talk will give insight into how varying levels can effectively communicate scientific and sustainable ideas to those in policy. Part of this talk will be describing the process for becoming a Science and Technology Policy Fellow of the American Association for the Advancement of Science, as well as the roles of a fellow in the Executive branch of the federal government and how sustainability can be communicated to an array of non-scientific audiences.

AUTHOR INDEX

Notes:

Notes:

Notes: